BEARING DESIGN
& FITTING

IAN BRADLEY

MODEL & ALLIED PUBLICATIONS
ARGUS BOOKS LIMITED
Station Road, Kings Langley
Herts.

ISBN 0 85242 463 9

Made and Printed in Great Britain by Unwin Brothers Ltd., Old Woking, Surrey

BEARING DESIGN & FITTING

PREFACE

ALTHOUGH the subject treated of in this book is complex and covers a very wide field, an attempt has, nevertheless, been made to deal with the main principles involved and, at the same time, to furnish examples of bearing design and application that may be found of use particularly in the small workshop.

At times, difficulties are bound to arise with regard to the application of ball or roller bearings to some particular mechanism, but, fortunately, leading manufacturers are always ready to offer expert advice in solving such problems.

We beg to acknowledge any references that have been made to articles appearing in the *Model Engineer* and, at the same time, we would thank Messrs. Ransome Hoffman Pollard Ltd., the British Timken Co., and the SKF Ball Bearing Co., for the information they have kindly supplied.

CONTENTS

CHAPTER   PAGE

One   PLAIN BEARINGS   .   .   .   1

Materials for bearings—Babbit metal—Bronze —Aluminium—Cast iron—Steel—Plastics— Design of bearings—Thrust bearings—Coned pivot bearings—Dimensions of bearings.

Two   PLAIN BEARINGS (*continued*)   .   .   14

Methods of bearing adjustment—Finishing bearing surfaces—Lapping—Running fits— Fitting one-piece bearing bushes—Split bushes —Relining bearings with white metal.

Three   PLAIN BEARINGS (*continued*)   .   .   32

Lubrication of bearings—Feeding oil to the bearing.

Four   BALL BEARINGS   39

Types of ball bearings—Single- and double-row—Self-aligning forms—Double-purpose bearings—Mounting ball bearings—Preloading—Protection of bearings—Ball thrust bearings—Single-row—Self-aligning—Double-row thrusts—Double thrust bearings—Mounting thrust bearings.

CONTENTS

CHAPTER   PAGE
Five   ROLLER BEARINGS   .   .   . 57
          Parallel roller bearings—Mounting bearings—
          Tapered roller bearings—Methods of mounting
          —Application to machine tools—Needle roller
          bearings—Lubrication of ball and roller bear-
          ings.

Chapter One

PLAIN BEARINGS

WHEN a rotating shaft is carried in one or more plain bearings, the most important requirements are that the bearing surfaces of the two components should be formed of materials that are highly resistant to wear, and, secondly, that the bearings as a whole should be designed to carry the load under all working conditions and with the minimum of friction.

In addition, the bearing surfaces should be highly finished, not only to lessen friction and wear, but also to assist in maintaining continuous lubrication, as well as reducing noise and vibration in working.

MATERIALS FOR BEARINGS

As a rule, the shafts and spindles of machines are made of steel in order to withstand working stresses and to provide against bearing wear. Mild steel may be used for this purpose, but alloy steels, containing chromium, vanadium and other metals, are preferable where the working conditions are more severe and good wearing qualities are demanded. Hardened steel shafts are highly resistant to wear, but the accurate finishing of the bearing surfaces is a more difficult matter and usually entails grinding and even lapping following the hardening process. An unhardened steel shaft will run satis-factorily in bearings made of a variety of materials, ranging from soft tin- or lead-base alloys to phosphor-bronze and cast iron.

BABBIT METAL

The comprehensive name for the numerous tin alloys used for bearings is Babbit metal or White metal. This material usually contains a large percentage of pure tin to which antimony and copper are added to increase its hardness and strength; a proportion of lead is often incorporated with beneficial results as regards both hardness and rigidity.

The high price of tin has led to an increased use of lead-base bearing metals, and good results are also being obtained with cadmium-zinc alloys in the bearings of aircraft engines.

These various alloys are used, for example, in motor-car crankshaft bearings and for lathe mandrel bearings with excellent results, as they are both durable and greatly reduce friction; by virtue of their softness they reduce shock and noise, and moreover, they will not score an expensive component such as a crankshaft, for should the lubrication fail, the bearing metal will melt at quite a low temperature leaving the shaft undamaged.

Increasing the proportion of antimony hardens the alloy and renders it more brittle, but this does not necessarily cause more rapid wear of the shaft, for bearing wear is usually more pronounced when the softer alloys are employed, owing to grit and abrasive metal particles becoming embedded in the material and acting as a lap in wearing the shaft.

Needless to say, these comparatively soft bearing alloys, which have but little inherent strength, require adequate support to carry the bearing load; they are, therefore, usually incorporated merely as a lining in a split bronze or steel shell rigidly backed by a steel or cast iron housing

As will be described later in detail, split bearing bushes are lined with white metal by first tinning their surfaces and then casting on the bearing metal in a suitable jig so that it adheres firmly and evenly. Formerly, a thickness of white metal of about $1/10$ in. was commonly used for bearings of moderate size, but much thinner layers of alloys, backed by thin steel strip, are now employed for the bearings of parts such as motor-car crankshafts and the big ends of connecting rods.

To afford a good working surface, the white metal lining may finally be accurately machined, preferably by means of a diamond tool.

Lathe mandrels are sometimes carried in split bushes formed of a special white-metal antifriction alloy having good wear-resisting properties as well as ability to withstand shock and heavy loading.

Babbit metals of many types are manufactured and sold under proprietary names; these vary in their composition so as to render them suitable for application over a wide range of ordinary as well as special requirements. When selecting a bearing metal for any particular purpose, it is therefore advisable to follow the manufacturers' recommendations as to the most suitable alloy to employ.

BRONZE

The name bronze is usually applied to alloys composed of copper and tin, as opposed to brass which generally consists of a mixture of copper and zinc.

The bronze used for bearings often contains additional substances to make it more suitable for its purpose; thus, a small amount of added phosphorus renders the material harder and tougher, besides improving the fineness of its structure. This alloy is known as phosphor-bronze, and it is, perhaps, the most widely-used bearing alloy.

Lead may be introduced to form a lead-bronze, which has greatly improved machinability, and, although the toughness of the material may thereby be reduced, its capacity to run without heating and to withstand seizure from under-lubrication is increased. When nickel is added to these bronzes, their strength and toughness are increased and, at the same time, any tendency to brittleness is greatly reduced. This alloy is largely used for highly-stressed parts such as worm wheels.

ALUMINIUM

This metal in its pure state is unsuitable for bearings, but in the form of duralumin, which contains small quantities of copper, magnesium, manganese, silicon and iron, it is on account of its lightness and strength sometimes used for making the connecting rods of small internal combustion engines, but adequate lubrication is, here, essential to prevent scoring of the bearing surfaces. Other alloys of aluminium are also used for bearings, but it is advisable to harden steel shafts running in aluminium alloy bearings where high speeds or heavy loading are employed.

CAST IRON

Cast iron is an excellent material for forming the bearings of even an unhardened steel shaft, and so is often used for the headstock mandrel bearings of small lathes and the spindles of drilling machines. After a little use, such bearings acquire a hard glazed surface which is highly resistant to wear; moreover, temporary under-lubrication does not damage the bearing surfaces as the graphite content of the cast iron itself serves to reduce friction and, at the same time, acts as a lubricant. Unlike most other metals, cast iron can be mated with cast iron to form a durable bearing; hence, the slides of machine tools are composed of this material and the slide gib pieces may also be made of cast iron.

STEEL

A soft steel shaft running under load in an unhardened steel bearing will not give satisfactory service, but, if these two components are hardened and their surfaces highly finished, a very durable form of bearing will result. The mandrel and its bearings in high-class precision lathes are constructed in this manner and, given proper lubrication, these maintain their accuracy indefinitely and will run for long periods without needing adjustment. The ordinary cycle chain is another case in point, and, where grit is excluded and constant lubrication maintained by means of a gear case, renewal is but seldom required.

Although there is no necessity to harden a steel shaft running in white metal bearings, it is advisable to do so when phosphor-bronze bushes are fitted and the bearing is highly-loaded or under-lubricated; for, if the gudgeon pin of an internal combustion engine is examined after a long period of use, it will be found that even the hardened pin itself is more worn than the connecting rod bush.

Manufacturers of machine tools have found that a highly-

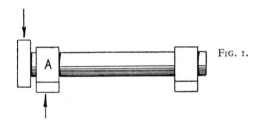

Fig. 1.

finished hardened steel spindle mounted in cast iron bearings will run for very long periods without requiring attention and, moreover, when the components are correctly designed the wear taking place is negligible.

PLASTIC MATERIALS

These substances are sometimes used for bearings designed to work under abnormal conditions; thus, the plastic known as Tufnol, in conjunction with a stainless steel shaft, provides an efficient bearing where corrosion is apt to occur as, for example, in mechanisms exposed to sea water. Under these conditions, the use of a non-metallic material for one component of the bearing eliminates local galvanic action.

DESIGN OF BEARINGS

Both the design of a bearing and the materials used in its construction depend largely on the duty it has to perform; the heavily-loaded, fast-running bearings of an internal combustion engine crankshaft must be correctly designed in every particular to give satisfactory service, whereas the requirements

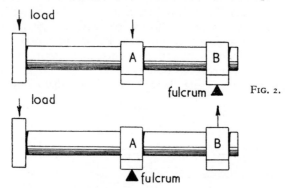

Fig. 2.

for the bearing of, say, a control lever are much less exacting and the latitude permissible in construction is correspondingly greater.

Nevertheless, there are certain essential principles in bearing design which apply in all cases, although the actual materials employed are sometimes of less significance; for example, in the small workshop bearings are commonly made of such proportions that they will safely carry loads greatly in excess of normal requirements.

The greatest difficulty is encountered where bearings of restricted size have to be designed to withstand heavy loading, as in aircraft engine construction.

As represented in Fig. 1, one bearing, A, should be positioned as close as possible to the point at which the load is applied, for if, as shown in Fig. 2, there is undue overhang and the bearing B is regarded as a fulcrum, then the pressure on the bearing A will be greatly increased. If, on the other hand, the bearing A is considered as the fulcrum, the pressure on B will then be greater than the load pressure owing to the

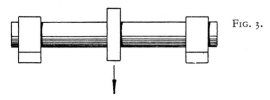

FIG. 3.

leverage exerted. Clearly, in either case the load pressure will be concentrated at the ends of both bearings, and under these conditions wear will be rapid so that the shaft will soon rock and touch only on the ends of the bearing bores.

If, as represented in Fig. 3, a shaft is carried in two widely-spaced bearings and the load is applied near the mid-point, some bending of the shaft may then occur, or an unbalanced component rotating at high speed may likewise deflect the shaft.

FIG. 4.

An example of this is an incorrectly installed lineshaft where the belt pull and unbalanced pulleys both tend to bend the shaft, thus causing rapid bearing wear.

The correct method, illustrated in Fig. 4, is to fit a bearing close to the point of loading; and should the load be heavy and the shaft liable to deflection, it is advisable to install a bearing on either side of a component such as a pulley, as shown in Fig. 5. The action of forces arising from unbalanced rotating and reciprocating parts is a problem encountered in the design of the crankshaft and its bearings in a four-cylinder, high-speed internal combustion engine. If the crankshaft bends under the influence of these forces, as it may well do when rotating at high speed, the additional load will fall mainly on the centre bearing of a three-bearing crankshaft and may be too great for the bearing to withstand.

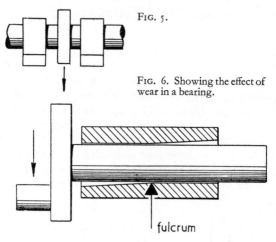

Fig. 5.

Fig. 6. Showing the effect of wear in a bearing.

fulcrum

To meet this difficulty, the crankshaft is made extremely stiff, otherwise, as experience shows, a third, centre bearing is of little practical value as at high speeds it becomes overloaded and, when worn, it ceases to support the crankshaft.

Where an overhung crankshaft is supported in a single long bearing, any wear taking place may cause a fulcrum to be formed in the bearing bore, about which the shaft can tip and thus cause bell-mouthing of both ends of the bearing.

This state of affairs is illustrated diagrammatically in Fig. 6, and, to overcome this failing, the bore of the bearing bush is recessed or chambered so that it makes contact with the shaft at the two outer thirds only of its length, as shown in Fig. 7.

As an alternative measure, the shaft itself may have its diameter reduced by a few thousandths of an inch over the middle third of its bearing surface, but as this slightly reduces the strength of the shaft it is usually preferable to recess the bush.

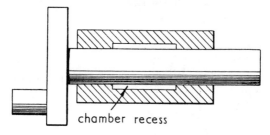

FIG. 7. Chambering the bearing to distribute the load.

A further advantage of this mode of construction, which actually provides two bearings accurately in line, is that an oil well is formed as an aid to the lubrication of the bearing.

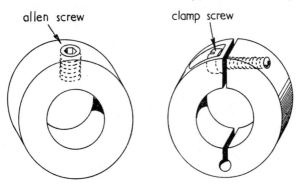

FIG. 8. Two types of thrust collars.

THRUST BEARINGS

Although the spindles of machine tools are nowadays usually fitted with ball- or roller-bearings to take the end-thrust, a plain collar secured to the shaft and butting against a bearing bush will be found adequate for lighter loads, or for use in the feed mechanisms of machine slides.

When installing lineshafting, the end-play, as well as the slight end-thrust sometimes occasioned by the belt drives, may

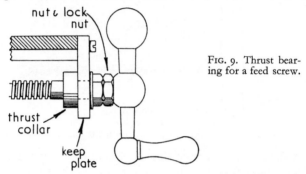

FIG. 9. Thrust bearing for a feed screw.

be taken up by shaft collars. Two patterns of these collars are illustrated in Fig. 8, and it should be noted that they have no projections which might catch in the clothing and endanger the operator.

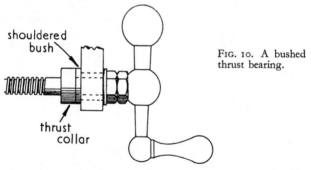

FIG. 10. A bushed thrust bearing.

Where the end-loading is heavier and accurate end-location is necessary, it is advisable that the thrust collar should be formed integrally with the spindle. A case in point is the feed screw operating a machine slide such as a lathe cross slide.

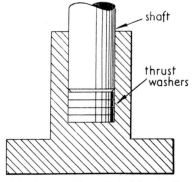

FIG. 11. A plain thrust bearing.

Here, as represented in Fig. 9, the integral thrust collar of the screw butts against the keep-plate secured to the slide, and the end-location and the reverse thrust are provided for by an adjusting nut fitted with a lock nut. As illustrated in Fig. 10, this mode of construction is sometimes elaborated by furnishing the keep-plate with a shouldered bush which must be firmly secured in the plate to prevent any possibility of end-wise movement.

The end-thrust in a vertical shaft may be taken in a foot-step bearing of the design shown in Fig. 11. Here, a series of washers consisting of alternate hardened steel and bronze discs is fitted.

This form of construction not only distributes the wear, but also allows the bearing to continue to work efficiently in the event of one pair of washers seizing owing to a temporary lack of lubrication.

CONED PIVOT BEARINGS

An example of this arrangement is seen when the 60 degree coned centre of the lathe tailstock is employed to support work mounted between the lathe centres. This form of bearing is designed to take both radial and thrust loads and is used in instrument work and sometimes to support the balance wheel shaft fitted to clocks. As illustrated in Fig. 12, the male cone centres may be formed either on the shaft or on the pivot screws. The pivot screws should have a fine thread to facilitate making exact adjustments, and lock nuts are fitted to maintain the setting. An alternative method of securing the pivot screws after adjustment is to slit the bearing housing, as shown in Fig. 13, so that it can be closed by means of a clamp screw to grip the pivot. Bearings of this form are more suitable for application to intermittent or rocking motion rather than continuous rotation.

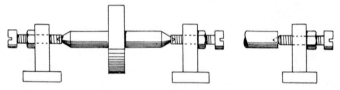

Fig. 12. Coned pivot bearings.

DIMENSIONS OF BEARINGS

Usually the diameter of the bearing journals, that is to say the portions of the shaft in contact with its bearings, is decided by fitting a shaft of sufficient stiffness and strength to carry the load imposed; the length of the bearing is then adjusted to distribute the load so that it does not exceed a safe maximum working pressure.

The essential feature of a lathe mandrel, for example, is that it should have a diameter sufficiently large to ensure adequate rigidity, and the length of its bearings is, in part, dependent on the requirements of design affecting the headstock as a whole.

The latter consideration and the nature of the bearing material impose some limitation on the efficiency of the bearings, and in consequence lathe manufacturers restrict the working speed of the mandrel. Where higher mandrel speeds are required it may be necessary to design the bearings specially for this purpose and, at the same time, to give the mandrel journals a particularly high finish. Precision lathes designed for high-speed running usually have a hardened-steel mandrel carried in hardened-steel bearings; moreover, these bearings are very highly and accurately finished, and the

FIG. 13. Method of secur-
ing the pivot screws.

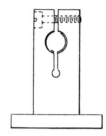

mandrel itself is of rigid construction to prevent any distortion under working conditions, which would result in damaging the bearing surfaces.

Bearing dimensions are also largely affected by the necessity of providing adequate lubrication, a question which will be discussed later; for the bearing may suffer rapid wear, or even seizure, if the oil film between the contact surfaces breaks down as a result of excessive pressure or unduly high speed.

Continuous running under constant load provides the least favourable working conditions as affecting lubrication, whereas lubrication is facilitated by either reversal of the load pressure or a partial rotation only, as in the case of a crank pin and a gudgeon pin bearing respectively.

These factors should, therefore, be taken into account when determining the length of the bearings suitable for any particular piece of mechanism; thus, the shaft bearings of an electric motor, designed for continuous running with a belt drive, are usually made of a length equal to some three times the shaft diameter. On the other hand, the crank pin of a stationary engine may have a bearing of the same length as the diameter of the pin. As a general rule, shaft bearings and those of machine tool spindles are made of a length equal to from one and a half to twice the shaft diameter.

Chapter Two

PLAIN BEARINGS
(CONTINUED)

As SOME wear, however small, must arise in all bearings, some means of adjustment must be provided to enable a plain bearing to maintain its efficiency.

The simplest method of providing adjustment in a bearing, such as the self-contained bearing in a cast iron lathe headstock, is to split the metal on one side only, as shown in Fig. 14, and to fit a screw for closing the bearing as a whole on the shaft.

This design allows only a small range of adjustment, for, if the cast iron is in this way bent excessively, it may easily be fractured along the line of flexion. To minimise wear and the consequent need of adjustment, it is advisable to employ a hardened mandrel when this type of bearing is fitted; in addition, the bearing cap should be closed against a laminated metal shim in order both to maintain rigidity and to prevent the escape of oil. This shim material usually consists of a metal strip, some ten thousandths of an inch in thickness, composed of layers each two ten-thousandths of an inch thick lightly held together by solder; this allows one or more layers to be easily removed to enable the bearing to be closed on the shaft with great accuracy.

It should be noted that, when the upper part or cap of the bearing is in this way only partly detached from the headstock casting, the alignment of the bearings is preserved and, at the same time, a solid face is presented to the mandrel for taking the thrust of the lathe tool.

The more usual construction to allow for taking up wear is to form the bearing in two parts with a detachable cap, as

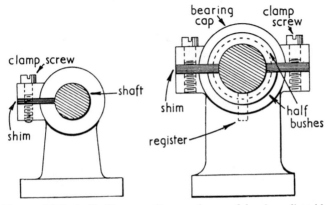

FIG. 14. Method of taking up wear in a split bearing.

FIG. 15. A capped bearing adjustable for wear.

represented in Fig. 15, the upper portion can then be closed on the shaft by the removal of the shim material. Bearings of this form are, as a rule, made with removable half-bushes, and these brasses or shells, as they are termed, are lined with a layer of Babbit anti-friction metal; but, as an alternative, the inset half-bearings may be made entirely of a suitable type of bearing white metal, as illustrated in Fig. 16. A register peg is sometimes fitted to prevent the bearing from turning in its housing, and the components should be clearly marked to ensure that after removal they are replaced in their original positions.

It will be clear that these methods of bearing adjustment operate in one direction only, and as a result of wear a shaft, such as a lathe mandrel, may eventually become out of alignment although afforded the correct bearing clearance. To obviate this, the adjustment of the bearing may be made to operate concentrically with the shaft. Some machine tools are equipped with spindles having a coned bearing surface as

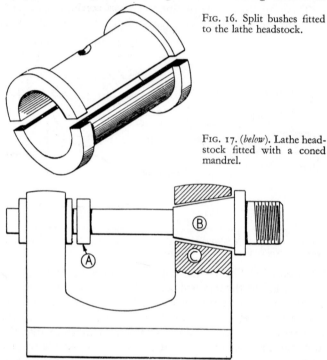

FIG. 16. Split bushes fitted to the lathe headstock.

FIG. 17. (*below*). Lathe headstock fitted with a coned mandrel.

depicted in Fig. 17; and to take up wear, the thrust collar (A) is adjusted to allow the spindle cone (B) to seat more deeply in its bearing (C).

Another form of concentric adjustment is provided, as in the Myford-Drummond lathe headstock, by closing a split, tapered bush on the parallel mandrel. Here, as shown in Fig. 18, the bush itself (A) is coned and is contracted by being drawn into the coned housing (B) by means of the threaded collar (C).

16

A screw (D), engages a pad piece fitted to the slot in the bush and thus keeps the bush expanded after adjustment has been made. This screw also prevents the bush from turning in its housing and, at the same time serves as a lubricator for the bearing. It should be noted that it is advisable to use a square thread for the adjusting mechanism, otherwise the act of tightening the collar may itself cause the end of the bush to close on the shaft.

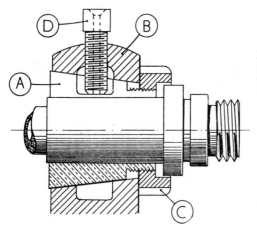

Fig. 18. Drummond type of lathe headstock bearing.

FINISHING BEARING SURFACES

The finish of the contact surfaces of a bearing not only affects its efficient working and wearing qualities, but is also important in promoting quiet and smooth running. If both the journal and its bearing are finished by ordinary turning and boring operations in the lathe, the surfaces so formed will, when highly magnified, have the appearance indicated diagrammatically in Fig. 19; that is to say, the surface is characterised by a series of ridges and hollows. It follows, therefore, that two such surfaces will make contact at only a number of points, and the total area of contact will be much reduced.

This means that pressure within the bearing will be concentrated on these points of contact, which will then tend to break through the film of lubricating oil, thus giving rise to heating and rapid wear of the bearing as a whole. Reference to Fig. 19 will make it clear that, when these high spots have been worn away, looseness in the fit of the bearing will develop, and although a bearing may appear, in the first instance, to be well

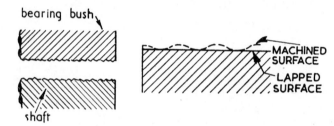

FIG. 19. Bearing surfaces finished by machining.

FIG. 20. Showing the effect of lapping a bearing surface.

designed and fitted, a close examination of the contact surfaces will often be a better guide to its future working efficiency.

In better-class work it is usual to finish the shaft by a grinding process and the bearing either by reaming, broaching or grinding. This results in a much smoother surface being formed on the bearing components, and the effect is represented diagrammatically in Fig. 20. Here, the remaining roughness is such that, after a period of running in, good contact will be established without sufficient metal being worn away to give rise to looseness.

When greater durability, together with quiet running, is desirable, the bearing surfaces of both the shaft and its bearing are usually finished by a lapping process, which will be described later. Commercially, this can be carried out in a machine specially designed for the purpose, but in the small workshop hand-lapping is usually employed, and, although this takes time and requires some skill, it is well worth while when the advantages of accurately fitted bearings are borne in mind.

Lapping is usually applied to hardened and unhardened steel shafts and cast iron and hardened-steel bearings, for soft bearing materials such as white metal are not suitable for treatment with ordinary lapping compounds of the carborundum type, as there is then a danger of the abrasive becoming embedded in the metal and so causing the bearing to act as a lap after it has been assembled.

Cast iron, on the other hand, can be lapped until the fine scratch lines are eliminated and the surface presents a mirror finish with no apparent roughness even under high magnification.

The spindles of drilling machines, which carry only a light radial load, when finished in this way will show negligible wear even after prolonged use; jockey pulleys, too, will run almost silently at high speeds and will not emit the roaring sound associated with roughly finished bearing surfaces.

When fitting some hardened steel rollers to their spindles by lapping, it was found that the components would spin freely when dry, but, when a little thin oil was introduced, this was no longer possible as the shearing of the extremely thin oil film between the accurately fitted bearing surfaces gave rise to sufficient friction to act as a brake.

When finishing bearing bushes made of phosphor-bronze or similar alloys, for example those fitted to the finely-ground shafts of electric motors, a turning operation employing a diamond-tipped tool is often used; this gives a high surface finish, and such bearings when adequately lubricated give satisfactory service for long periods even when run under heavy continuous load.

LAPPING BEARINGS

Lapping is a process whereby shafts and bearings are finished to the exact diameter required and, at the same time, they are rendered truly circular and parallel. This is accomplished by means of a cylindrical or hollow-cylindrical tool, known as a lap, charged with an abrasive compound.

For lapping a shaft externally, a tool of the form illustrated in Fig. 21 is commonly used. The lap is usually made of cast

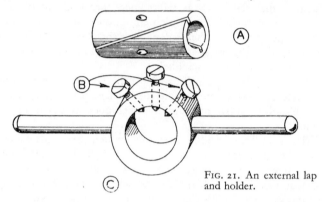

FIG. 21. An external lap and holder.

iron when used for lapping steel shafts, but softer metals like copper, aluminium and lead are also employed although they wear more rapidly during the lapping process.

It will be seen that the lap itself (A) is split longitudinally so that, when the screws (B) of the lap holder (C) are tightened, the lap can be closed on the shaft to take up wear; where the lap has thick walls, it is advisable to increase its flexibility by cutting a slot opposite the oblique slit.

Manufacturers of abrasives, notably the Carborundum Co., now make compounds suitable for lapping all bearing metals in common use, and it is advisable to follow the manufacturers' recommendations when selecting a lapping compound for any particular class of bearing. It is stated that, when hand-scraping is employed, effective contact in bronze or white metal bearings rarely exceeds 20 per cent, whereas lapping may produce contact between the shaft and its bearing as high as 80 per cent of the total bearing area.

In external lapping practice, the shaft is rotated at a medium speed and the lap, lightly charged with abrasive, is moved to and fro along the shaft. From time to time, the lap is cleaned and then readjusted so that it continues to make light frictional contact with the work. By employing a series of lapping compounds, and beginning with one of coarse grain, the process will be more quickly carried out, and the fine-grain abrasive finally used will impart a high finish to the work.

For the internal lapping of bearings and bushes, a cast iron lap of the form shown in Fig. 22 will give good results. Here,

FIG. 22. An internal expanding lap.

the lap itself, is also split but is formed with a taper bore. To counteract wear, the lap is expanded by being pushed further on to the tapered mandrel on which it is mounted. Exact adjustments will be more readily made if the end of the mandrel is threaded and a nut is used to push the lap forward. The Boyar-Schultz copper-head lap, illustrated in Fig. 23, is a very convenient form of a commercially produced tool; three copper sheaths are supplied with each lap holder to enable different grades of abrasives to be used separately. In this device the lap proper is made of sheet copper, formed to the exact nominal diameter and carried on a split mandrel; this mandrel can be expanded as required by tightening the adjusting screw shown.

As lapping is employed as a means of truing a part and eliminating tool marks whilst bringing it to the finished diameter, it is advisable to leave only a small amount of metal, say half a thousandth of an inch, to be removed in this way, for lapping is a somewhat slow process and time will be wasted if this allowance is unnecessarily great.

As already mentioned, soft metals such as bronze and white metal can be successfully lapped to a good finish, but for this purpose it is essential to use the correct type of abrasive, otherwise the lapping compound, becoming embedded in the metal, will cause damage to the bearing surfaces under working conditions.

ADJUSTING
SCREW

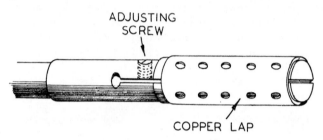

COPPER LAP

FIG. 23. The Boyar-Schultz internal lap.

Both hardened and unhardened steel shafts can be lapped with fine emery or carborundum powder mixed with thin oil, and the surface is brought to a high finish by employing either a fine grain abrasive of this type or by using one of the compounds specially manufactured for this purpose; as an alternative, powdered oilstone will give good results when finish-lapping either hard or soft bearings, for it has little tendency to become embedded in the metal.

Whatever lapping compounds are used, it is essential to clean the bearings thoroughly with paraffin before they are assembled, otherwise any remaining abrasive will cause rapid wear.

RUNNING FITS IN BEARINGS

In the example cited of the closely-fitted spindle bearings of a drilling machine, the bearings are only lightly loaded, although the speed may be high, and consequently heating is not apt to occur. When, however, the bearings have to carry a heavy load, it is essential to provide sufficient working clearance between the contact surfaces to ensure that the running friction does not cause heating of the bearings; for this may give rise to breakdown of the oil film and damage to the bearing surfaces.

In the small workshop the usual practice when fitting bearings is, as previously mentioned, to ream, or ream and lap, the bearing in order to make the bore circular and parallel and, at the same time, to give a good finish to the working surface; the shaft is then machined and lapped to fit the bearing. However, to enable the assembled bearing to operate satisfactorily, there must be some clearance between the two components to afford space for the film of oil that should separate the working surfaces.

This clearance, also termed the fitting allowance, is kept very small in well-fitted bearings in order to prevent shake and to maintain the working parts in true alignment. Where the bearings are fitted individually, the fitting process is continued until the parts are found to work smoothly together and with sufficient freedom to guard against scoring or seizure under working conditions. Experience alone will enable the worker to recognise the feel of a correctly and closely fitted bearing, and if such parts can be measured with the aid of the micrometer this will serve as a guide for future work.

It should be borne in mind that a bearing assembly consisting of a steel shaft running in a cast iron bearing can be fitted very closely without risk of damage, for cast iron contains graphite which serves as a lubricant and, in addition, a hard, glazed surface highly resistant to scoring is formed on this material under working conditions. Phosphor-bronze bushes, on the other hand, usually require rather more working clearance as, otherwise, a temporary failure of lubrication may result in damage to the bearing surface of the steel shaft.

Clearly, hand-fitting operations such as these, requiring both skill and time, cannot be undertaken profitably in ordinary commercial work; nevertheless, they are essential where precision fitting is called for in instrument making and other like trades.

In commercial engineering there are commonly two essential requirements: rapid assembly of components in the first instance, and the production of spare parts that can be installed without a preliminary fitting operation. As a shaft and its bearings, for example, are separately manufactured and may even be made in different factories, it is essential that some method of standardization should be adopted to ensure interchangeability of parts; furthermore, components can only be machined by manufacturing methods to a certain degree of accuracy. In view of these considerations, standards are recommended imposing definite high and low limits to both the diameter of shafts and the bore of bearings; the difference between these two limits is known as the tolerance. Tolerances are so arranged that, should the largest permissible shaft be fitted to the smallest bearing, sufficient clearance will remain to ensure efficient working.

An example taken from the recommendations of the Engineering Standards Committee of Great Britain will, perhaps, help to make the matter clear. In first-class work, the maximum diameter allowed for a 1 in. shaft is 1.000 in. and the minimum 0.999 in., giving a difference or tolerance of 0.001 in. As the smallest bearing is allowed a bore of 1.001 in., the largest shaft will then be afforded a working clearance equal to 0.001 in.

On the other hand, as the smallest shaft diameter is 0.999 in. and the maximum bearing bore 1.002 in., the clearance in this instance is 0.003 in. Although sufficient working clearance is, of course, essential, excessive clearance may in some bearings be highly undesirable, and methods are therefore adopted to reduce the maximum clearance that may occur when a standard system is in use. In the first place, manufacturing practice aims at machining parts as far as possible to dimensions midway between the upper and lower limits imposed by the system. Greater accuracy in the fit of components can also be obtained by adopting a system of what is known as selective assembly. Here, the components are sorted into, say, three categories, large, medium and small; this enables a large shaft to be paired with a large bearing and, likewise, a small shaft with a small bearing.

FITTING ONE-PIECE BEARING BUSHES

Material for making bushes can be obtained in the form of cast rods consisting of phosphor-bronze or cast iron, but much less machining will be required if bronze rod is bought already turned and bored to fine limits of accuracy. The Glacier Metal Co. produce both solid and bored lead-bronze bars in 7-in. lengths; the solid rods are made in diameters ranging from $\frac{5}{8}$ in. to 6 in. and having an outside dimensional limit of from two to six thousandths of an inch. The smallest hollow bars have an outside diameter of 1 in. and a bore of $\frac{1}{2}$ in. and are machined to fine limits of accuracy both internally and externally.

This material is easily machined and has good wearing qualities, even when lubrication is somewhat scanty and the shaft is composed of unhardened steel. The manufacturers recommend that a bearing clearance of from one to one and a half thousandths of an inch per inch of the shaft diameter should be allowed when fitting.

Bushes machined from this material should be made a firm press-fit in their housings, that is to say, the diameter of the bush should exceed that of the housing by approximately one and a half thousandths of an inch per inch of the housing bore.

Bushes should always be pressed and not hammered into place; this can be effected either by pressing the parts in the vice between hard wood or aluminium clams, or by using an ordinary bolt and nut to supply the necessary pressure.

The entering end of the bush should be lightly chamfered, and care must be taken to ensure that the pressure is applied evenly and truly in line with the axis of the bush.

When a bush is pressed into its housing in this way it will

become slightly contracted, and the amount the bore diameter is reduced will depend on the rigidity of the housing.

After insertion, therefore, the bush should be again sized with the reamer to accommodate the shaft.

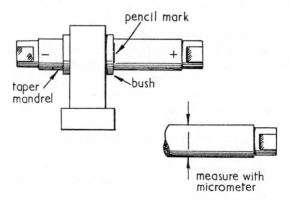

Fig. 24. Measuring the exact bore diameter of a bush.

An easy method of measuring the bore of a bearing is to insert a hardened-steel tapered mandrel of the kind used for mounting work between the lathe centres; as shown in Fig. 24, a pencil line is drawn on the mandrel to mark its depth of entry and a micrometer measurement is taken opposite this line. Moreover, this test, if repeated at the other end of the bush, will reveal any tapering of the bore.

When fitting bearings, it is advisable first to fit the bush in place and then to measure its bore in the manner described; following this, the shaft can be finished to the correct diameter to allow the necessary working clearance.

An alternative method of fitting bearings is to use ready-made bushes of Oilite manufacture. These bronze and lead-bronze bushes are impregnated with oil so that they remain self-lubricating for long periods. Although the bore of these bushes can be reamed to size after fitting, it is preferable not to do so but rather to finish the shaft to fit the bearing when in place.

The Manganese Bronze and Brass Co., of Ipswich, who manufacture these bushes, issue a booklet giving explicit instructions for fitting and maintaining bearings of this type.

SPLIT BUSHES

Bushes of the type fitted to the big end of a connecting rod working between the crank webs are made in two equal halves; this enables the rod to be assembled on the crankshaft and also provides for taking up wear.

The two half-bushes, forming the base and cap portions of the bearing, may consist either of bronze alone or of a steel or bronze shell lined with white metal. If shims are fitted between the two parts of the bush, wear can be taken up by removing one or more of the metal leaves, but, where the wear is unevenly distributed, the bush must be again bedded to the shaft by hand-scraping the contact surface. When a split bearing lined with white metal has become much worn, the usual practice is to re-metal the bearing; the white metal lining is then machined and finally scraped to fit the shaft.

In the motor-car industry strip-steel shells faced with a thin layer of white metal are now largely used for the main engine bearings; these are made to precision limits so that replacements, when needed to overcome wear, are easily carried out without dismantling the whole engine. When refitting bearings, on no account should the bearing cap with its half-bush be filed down to allow it to close on the shaft, for this destroys the circularity of the housing and renders subsequent fitting much more difficult.

To refit the white metal lined bearings of a crankshaft, the crankcase is supported base uppermost and the bearing caps are removed. The shaft is then bedded evenly in the base portions of the bearings by a process of hand-scraping, aided by the application of marking paste to indicate the high spots; when satisfactory bearing contact has been obtained, the bear-

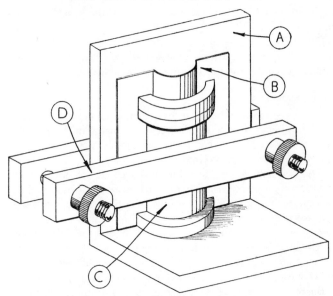

FIG. 25. A jig for remetalling bearings.

ing caps are bolted down and they too are scraped in until proper contact with the shaft journals is established.

As previously mentioned, the mandrels of some lathes run in split bushes entirely composed of a special brand of white metal adapted to resist wear and withstand shock. Bearings of this type are usually provided with metal shims to allow adjustment to be made with great accuracy and without danger of upsetting the bearing alignment. A further advantage of this form of bearing, in contrast with phosphor-bronze bushes, is that, even in the event of under-lubrication, the mandrel itself does not become scored and the damaged half-bushes can be readily and cheaply replaced.

RELINING BEARINGS WITH WHITE METAL

Reference has been made to the use of white metal linings

for the split bronze bushes fitted to engine crankshafts and connecting rod big end bearings.

In time, as a result of wear or damage, these linings will require renewal in order to restore the efficiency of the bearings. Needless to say, experience and skill are required to carry out this work properly, and as there are many different brands of white metal available, the manufacturers' instructions should always be closely followed in order to obtain the best results with the particular bearing metal used. Nevertheless, the process is essentially similar for all white metals, but details, such as the temperature required for casting, will of course vary. In the first place, the old metal must be run out by heating the bushes in a gas oven or with a blowpipe; the shells are then thoroughly cleaned and their inner surfaces roughened with a file to afford a hold for the new metal. Next, these surfaces are tinned with either pure tin or tinman's solder, using killed spirits as a flux and heating the back of each bearing evenly with a blowpipe.

While the bearings are still hot, the surplus metal is wiped away. In order to cast a layer of white metal on to the bearing surfaces, each half-bush must be held in a suitable jig, and Fig. 25 shows the type of jig commonly used for this purpose. The body (A) of the jig consists of an iron angle plate to which is attached a semicircular sheet-iron core (B), whilst the bush itself is held in place by the clamp (D).

The white metal is melted in an iron pot by applying a steady heat such as that obtained from a gas ring.

It is very important not to overheat the metal, as this will render it brittle and unfit for use. Manufacturers state the temperature to which their products should be heated and, in addition, they usually describe simple tests for ascertaining the temperature of the molten metal. One such test is to push a spill of white paper into the melted metal, and the correct pouring temperature is indicated by the paper turning brown, but should the paper catch alight, the temperature is too high. The molten metal should be kept well-stirred, and when the

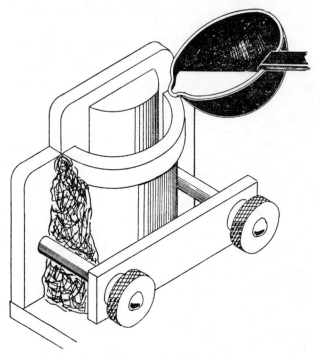

Fig. 26. Pouring the White Metal.

correct temperature has been reached the metal is poured into
the jig in the manner represented in Fig. 26; but, to prepare
the bearing for the white metal, the blowpipe flame is directed
on to the jig until the surface tinning on the bearing melts.
As soon as the metal has been poured into the jig, a venting
rod, made from a length of clean steel wire, is dipped in and
out of the molten metal in order to release any imprisoned air
and so allow the space between the jig and the bearing to
become completely filled. The metal should then be cooled
as quickly as possible, either by means of a blast of air from

the blowpipe or by applying a wet rag to the lower part of the bearing. To continue the work on the bearing, the two halves are next mounted together in the lathe, by employing a V angle-plate or other fixture, and the bore of the bearing is machined to size.

This machining operation is followed by hand-scraping the bearings to their shaft in the manner previously described.

Chapter Three

PLAIN BEARINGS
(CONTINUED)

THE adequate lubrication of a plain bearing is dependent on the maintenance of an oil film between the contact surfaces. As the shaft rotates, a wedge of oil under pressure is built up within the bearing, and it is this oil wedge that keeps the bearing surfaces apart so that metal to metal contact does not occur.

Should the bearings supporting a shaft be out of line, the bearing pressure may then be localised and concentrated at one or more points; this may result in the oil film being penetrated in these situations and, where metal to metal contact takes place, heating and scoring of the affected surfaces will, under conditions of heavy loading or high speed cause bearing damage or seizure.

Even if the bearings are correctly aligned, the high spots and ridges found in roughly-finished bearings are themselves apt to penetrate the oil film with like results. The separation of metal particles from the bearing surfaces, or the entry of dirt or swarf, may also cause local breakdown of the oil film leading to general bearing failure. The possibility of failure, due to misalignment of the bearings, is somewhat lessened by the use of the softer bearing materials, which also allow particles of grit or metal to become embedded in their surface and thus rendered less harmful to the oil film.

As to the type of lubricant used, the oil must have sufficient viscosity under the prevailing working conditions to ensure separation of the bearing surfaces. Too heavy a grade of oil applied to a fast running bearing will result in heating and thinning of the lubricant, attended possibly by bearing damage.

Heavy bearing loads and low speeds, however, call for a heavy oil, but, where high speeds are concerned, a best-quality lubricant of lower viscosity should be employed.

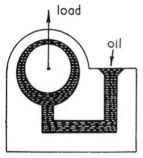

Fig. 27. Correct form of bearing lubrication.

Fig. 28. Incorrectly-designed oil feed.

FEEDING OIL TO THE BEARING

The first rule is that the oil should be supplied to the bearing at the point of least shaft pressure, as illustrated in Fig. 27; the adhesiveness of the oil to the shaft then causes the lubricant to be pulled into the bearing and there form a wedge between the metal surfaces. If, on the other hand, as shown in Fig. 28, the oil is supplied to the pressure side of the bearing, the shaft will tend to seal the oil-way and also the lubricant will not readily be drawn between the bearing surfaces.

The bushes of long bearings are generally provided with oil grooves for the purpose of distributing the lubricant, but these grooves should not be formed in the pressure area for in that situation they are not usually effective and, in addition, they reduce the bearing area where it is most required.

Oil grooves should not be cut to the full length of the bush, as this would tend to allow the oil to escape from the bearing.

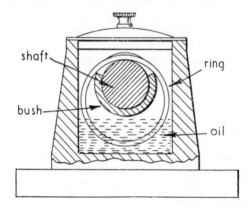

FIG. 29. Bearing fitted with ring-lubrication.

Ring-oiling is a useful method of maintaining a constant supply of oil to a rotating shaft. Here, as shown in Fig. 29, a loose metal ring is carried on the shaft in a gap formed in the bearing bush. The ring dips into an oil reservoir and, as it is rotated by the revolving shaft, an abundant and continuous supply of oil is carried up to the bearing. The edges of the gap formed in the bush should be relieved to facilitate entry of the oil between the bearing surfaces, for a sharp edge in this situation would tend to scrape off the oil from the shaft.

This oiling system needs but little attention, and it is only necessary to maintain the oil level to cover the lower part of the ring, and to clean out the reservoir at long intervals.

A method, now widely used for ensuring a continuous supply of oil to the bearings of small electric motors, is to provide an oil reservoir surrounding the bearing and to form a gap in the upper surface of the bush; the reservoir is packed with wool yarn, maintained in contact with the exposed area

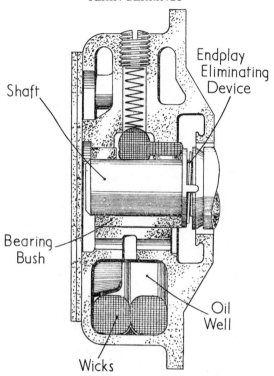

Shaft

Endplay
Eliminating
Device

Bearing
Bush

Oil
Well

Wicks

FIG. 30. Lubrication system of the Crompton Parkinson electric motor.

of the shaft by means of a light spring fitted to the oiling nipple.

An example of this oiling system is illustrated in Fig. 30, which represents the construction of the bearings fitted to the Crompton Parkinson fractional horsepower electric motors. The manufacturers claim that not only is this system of lubrication fully efficient, but also replenishment of the oil is necessary only at long intervals.

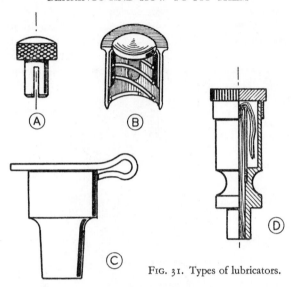

FIG. 31. Types of lubricators.

Continuous lubrication of bearings can also be effected by the use of Oilite self-lubricating bronze bearing bushes. This material, by virtue of its special composition, contains a large percentage of oil which ensures satisfactory lubrication usually for as long as the life of the bush. The secondary shafts and control mechanisms of machine tools are now largely carried in bearings of this type, and, where such shafts run at high speed, supplementary lubrication may be provided by supplying oil in the ordinary way through a lubricator. For this purpose, no oil-way need be formed in the bush itself, for the porosity of the material will enable the oil to soak through the bearing.

Where, as in some manufacturing processes, the presence of oil is objectionable, bushes impregnated with graphite to serve as a lubricant are sometimes used.

The lubrication systems so far described are self-contained

36

and require replenishment only at long intervals, but the more common designs are dependent on the frequent addition of small quantities of oil to make good the loss from the bearings.

The simplest and most primitive way of lubricating a bearing is to provide an oil-way leading through the bearing

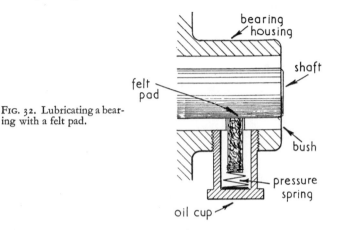

FIG. 32. Lubricating a bearing with a felt pad.

bush to the shaft, but in order to keep out dirt and chips it is advisable to close the hole with an easily removable plug of the type illustrated in Fig. 31A. The flush-fitting Bennet lubricator shown in Fig. B is neat in appearance and closes automatically after the oil can has been applied. The cycle-type lubricator, Fig. C, fitted with a spring-controlled, hinged lid, has the merit of providing a small oil reservoir and, at the same time, ensures that no dirt is carried into the bearing when the oil can is used.

The oil cup illustrated in Fig. D is useful for fitting to lathe mandrel bearings and to machine spindles generally, for the wick by siphonage action delivers a slow but continuous supply of oil to the bearing. Oil cups are sometimes fitted in the inverted position, as illustrated in Fig. 32, and the lubricant is then carried through a hole in the bush to the shaft by means of a felt pad, maintained in contact with the shaft by a light spring.

This form of lubrication is applied to the spindle bearings of small electric motors, but replenishment of the oil is required more often than when a large wool-packed reservoir is employed for maintaining the oil supply to the bearings.

Chapter Four

BALL BEARINGS

BALL bearings have many advantages over plain or sliding bearings: starting friction is much reduced, and loss of power when running under load is less owing to the lower coefficient of friction between the bearing surfaces. Lubrication, too, is greatly simplified, and the efficiency of the bearing is independent of the viscosity or temperature of the oil.

A ball bearing is also more compact than a plain bearing and wear is extremely slow; moreover, when renewal becomes necessary the bearing is readily replaced and no preliminary bedding in of the bearing surfaces is required.

The races of ball bearings are precision ground to very fine limits of size, and the actual ball tracks are finished by a lapping process to remove all grinding marks. The balls themselves are finished to within one ten-thousandth of an inch as regards both sphericity and overall diameter.

The diametrical clearance or radial play in an assembled ball bearing, as represented in Fig. 33, varies according to the grade of the bearing; thus, where it is necessary for the spindle of a machine tool to run with great accuracy, a bearing of Grade O will have a diametrical clearance of one ten-thousandth of an inch only. A grade OO bearing has a rather greater clearance to allow for inaccuracies in the machining of the bearing mounting, and Grade OOO, with even greater clearance, will permit the inner race to be slightly expanded when working conditions require that this component should be a force fit on the shaft.

These three grades of bearings have, in addition, some axial play or end float, amounting to approximately one and a half

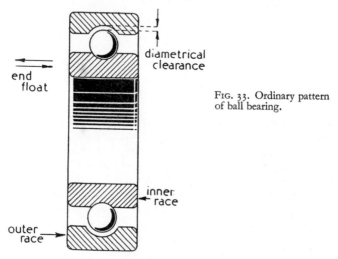

FIG. 33. Ordinary pattern of ball bearing.

ten-thousandths in the Grade O and five then-thousandths of an inch in the Grade OOO bearing.

The balls may be fitted in the bearing either in the form of a continuous circle, in which case the bearing is said to be filled, or they may be separated by means of a bronze, pressed steel, or plastic cage. This cage prevents the opposed surfaces of the balls from rubbing against one another, but the cageless type of bearing will carry a rather greater load owing to the increased area of contact between the balls and the bearing races.

TYPES OF BALL BEARINGS

Single- and Double-row Bearings. The type of ball bearing most commonly used is that illustrated in Fig. 34A; this has a single row of balls carried in a cage. These bearings are designed primarily to carry radial loads, but they are also capable of taking some thrust load. The light pattern are made with bores of from $\frac{1}{2}$ in. upwards, whilst the smallest metric size has a bore of 6 mm. The light, narrow type bearing, depicted in Fig. B, in the smallest size has a bore of $\frac{1}{4}$ in., or 10 mm. in the metric series, and the width of the former is only $\frac{7}{32}$ in.

These narrow, compact bearings will be found most useful for carrying light loads, when fitted to models and other small mechanisms, where good wearing qualities are an advantage and space is limited. The medium and heavy patterns of the ordinary type bearings have minimum bore sizes ranging from $\frac{1}{4}$ in. to 17 mm.

For carrying heavier loads, the bearings illustrated in Fig. C, having a double row of balls, are made in metric sizes of the light, medium and heavy pattern, the smallest bore size being 10 mm. in the first two types and 17 mm. in the third.

SELF-ALIGNING BEARINGS

Where a single row of balls is fitted, as shown in Fig. D, the outer race has its external surface formed as a portion of a sphere to fit into a housing of corresponding shape. This allows the bearing as a whole to tilt and accommodate itself to a shaft running out of line. In addition to carrying radial loads, these bearings will withstand end-thrust.

The compact form of self-aligning bearing illustrated in Fig. E is more commonly used. Here, a double row of balls is fitted, and the spherical form of the outer ball track enables the inner race, together with the balls, to align itself in the bearing and so compensate for any misalignment of the shaft.

This type of bearing will also withstand end-thrust, such as occurs in line shafting, but if the thrust is excessive, one row of balls may become unloaded and the resulting ball spin will cause damage to the bearing itself.

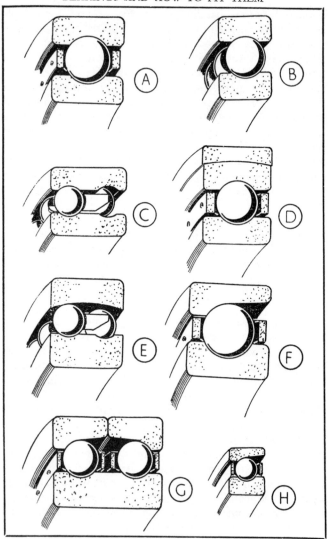

Fig. 34. Various types of ball bearings.

As shown in Fig. 34 F, these bearings have the ball track of the outer race formed to take end-thrust in one direction, and at the same time journal or radial loads are carried as in the ordinary type of bearing.

Where, as depicted in Fig. G, a double row of balls is fitted to a bearing of this type, the outer race is made in two parts, which must be securely clamped together when the bearing is assembled in place in order to allow the end-thrust to be taken in either direction.

The magneto-type bearing, illustrated in Fig. H, is designed to carry light radial and thrust loads. It is of similar construction to the single-row double-purpose bearing, but additional features are that the outer race is detachable and the bearing components are interchangeable. The smallest bearing of this pattern has a bore of 5 mm.

FITTING BALL JOURNAL BEARINGS

In the first place, it is essential to maintain strict cleanliness when handling ball bearings, for any foreign matter which gains entry at the time of fitting will do serious damage to the working surfaces after the bearing has been put into use.

To enable the bearing to be properly fitted and so to give satisfactory service, the surfaces of both the shaft and the bearing housing must be accurately machined to a high finish.

The revolving race, which is usually the inner race, must be made a light press fit in order to prevent it from creeping on the shaft when under load, but if excessive pressure has to be used to fit the race in place, it may be expanded and distorted so that the necessary diametrical clearance, already mentioned, is abolished; this will cause hard irregular running and the bearing as a whole will rapidly deteriorate in use.

Whenever possible, therefore, it is preferable to adopt the alternative method of securing the race to the shaft by means of a clamping screw or nut, as shown in Fig. 35. The shoulder against which the race abuts must also be machined flat and truly at right angles to the shaft; moreover, as illustrated in the drawing, the height of this shoulder should be sufficient to allow it to stand clear of the radius formed on the end of the race.

The stationary race, which in the example considered is the outer race, should be made a sliding fit in its housing, and where the bearing also serves to end-locate the shaft, this race too is clamped in place in the manner illustrated in Fig. 35.

The clamping collar used for this purpose may either be pressed inwards by means of screws or, as depicted, the collar itself can be made to screw into the housing.

When a second ball bearing is fitted to the shaft, either its inner or outer race must be left free to slide axially so that, under working conditions, the mounting itself imposes no end-thrust on the bearings. The usual methods of fitting are illustrated in Fig. 36. In Fig. A it will be seen that although both races of the left-hand bearing are clamped in place, the

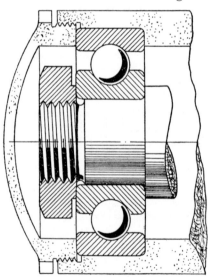

FIG. 35. Method of locating the bearing races.

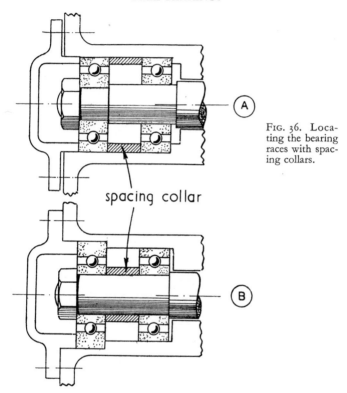

spacing collar

Fig. 36. Locating the bearing races with spacing collars.

inner race of the right-hand bearing is free to slide and align itself with its outer race. In the alternative method shown in Fig. B the spacing collar is fitted between the two inner races, thus leaving the outer race of the second bearing free to slide axially in its housing

Where the double-purpose bearings illustrated in Fig. 34F are used, as shown in Figs. 37A and B, to carry both end-thrust and radial load, either packing shims are inserted for adjusting

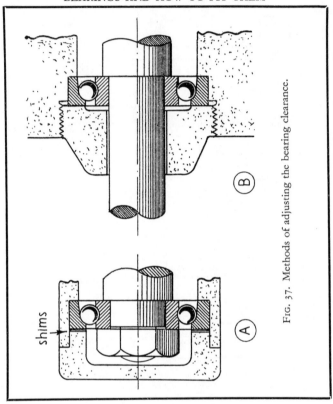

Fig. 37. Methods of adjusting the bearing clearance.

the bearing to take up end-play, or a screwed collar, secured by a locking screw or nut, can be used for this purpose.

An alternative method of fitting these bearings to take end-thrust in both directions is to mount two bearings either face to face or back to back in a single housing, as shown in Fig. 38.

The magneto-type bearing, Fig. 34H, is used in the same way in light mechanisms to provide for end-location and, at the same time, to take both radial and thrust loads.

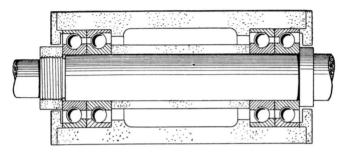

FIG. 38. Shaft mounting designed to take end-thrust in either direction.

adjusting collar

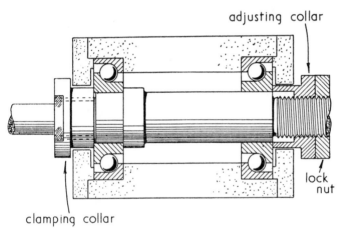

lock
nut

clamping collar

FIG. 39. Magneto-type bearings applied to a machine spindle.

Fig. 39 illustrates the manner in which a pair of magneto bearings was applied to a small drilling and grinding spindle designed for attachment to the lathe tool post.

Here, both outer races are retained in place by means of clamping collars; one inner race is clamped against a shoulder formed on the spindle, but the other is made an accurate slid-

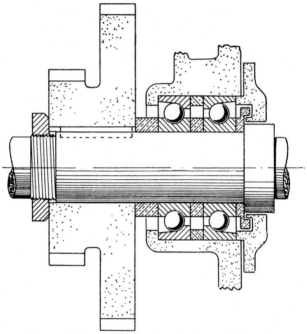

FIG. 40. Method of pre-loading a lathe mandrel bearing.

ing fit on the shaft so that end-play can be taken up by adjusting the threaded collar.

REMOVING BEARINGS

When designing mountings for ball bearings it is advisable to make provision for the removal of the bearings as a whole or their separate races. Races which are retained by means of clamping screws or nuts need not be tightly fitted, and their removal should present no difficulty, but where a race is made a press fit either on its shaft or in a housing, a portion of the race should be allowed to project in order to provide a shoulder against which pressure can be brought to bear.

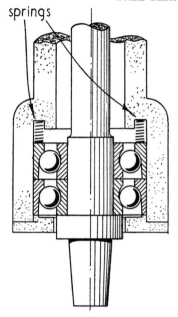

springs

FIG. 41. Pre-loading bearings by means of spring pressure.

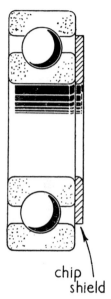

chip shield

FIG. 42. A bearing chip guard.

The method shown in Fig. 35 of fitting the inner race to its shaft illustrates this point.

Ball races can be withdrawn with a small wheel puller or by using other applications of screw pressure. Inner races, if well supported, can be driven off with light hammer blows carefully applied to the end of the shaft, whilst outer races can usually be driven out by using a well-fitting drift.

Hammer blows must never be applied to the race through the balls themselves, as this will form indentations in the ball tracks and so ruin the bearing.

PRELOADING BEARINGS

Where special precision-type ball bearings are fitted to the

49

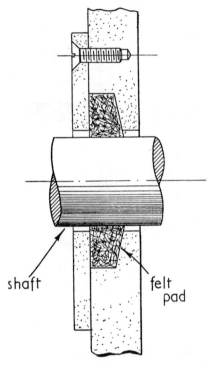

Fig. 43. Protecting the bearing with a felt washer.

shaft

felt pad

spindles of high-class machine tools, all trace of end-play can be eliminated by applying a permanent fixed load to the bearings at the time of assembly.

This preloading, as it is termed, may be effected in the manner represented in Fig. 40, which shows two double-purpose bearings mounted back to back to carry the front end of a lathe mandrel; the preload is applied by clamping the inner and outer races against accurately-fitted distance collars, which afford the correct amount of separation between the outer races to remove all end-play.

The bearings of small, high-speed spindles are sometimes

preloaded by fitting a series of springs to compress the outer races against the balls; the method employed is represented diagrammatically in Fig. 41.

PROTECTION OF BEARINGS

Just as it is essential to keep ball bearings clean during fitting, so it is equally important to prevent the entry of chips and other foreign material under working conditions. Whenever possible, the bearing should be protected by means of an end-cap or plate, such as is usually fitted to the non-driving end of an electric motor shaft and is illustrated in Fig. 35. But, where the driving shaft projects beyond the bearing, a flat, circular shield of the form shown in Fig. 42 may be fitted; here, the shield is attached to the rotating shaft and the edge of the disc is kept just clear of the outer ball race.

Felt ring pads fitted to the bearing housing and making light rubbing contact with the shaft afford an effective means not only of excluding dirt, but also of maintaining the lubricant within the bearing. Fig. 43 illustrates the method of fitting a pad of this sort to protect a shaft bearing.

Annular grease grooves of the form illustrated in Fig. 44A are fitted with a minimum of running clearance as an alternative means of retaining the lubricant, but this arrangement is more effective if used in combination with a felt pad, as shown in Fig. 44B. The labyrinth form of grease seal acts well in practice, and an example of this device in common use is illustrated in Fig. 45.

BALL THRUST BEARINGS

Single-row Thrusts. As shown in Fig. 46A, these have a single row of balls carried in a perforated cage and the two races are formed with curved ball tracks. Light, medium and heavy duty types are available, and the smallest bore size is $\frac{1}{4}$ in., or 10 mm. in the metric series. The form of thrust bearing shown in Fig. 46B has flat ball tracks and is intended for light loads only; $\frac{1}{4}$ in. is, here, the smallest bore size obtainable.

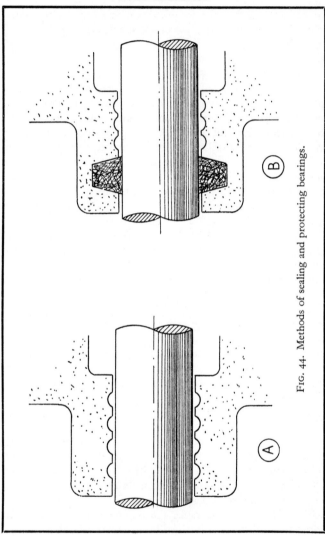

Fig. 44. Methods of sealing and protecting bearings.

Self-aligning Thrust Bearings. These bearings are used where the shaft is not correctly aligned with the thrust face.

As illustrated in Fig. 46c, one race is formed with a spherical surface where it abuts against a corresponding annular seating.

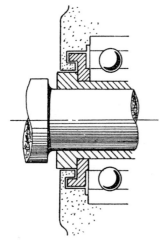

FIG. 45. Labyrinth form of grease seal.

The bore sizes of these bearings are similar to those of the preceding types.

Double-row Thrusts. Bearings of the type illustrated in Fig. 46D are designed for carrying exceptionally heavy thrust loads, and are sometimes made with one race in the form of two concentric rings in order to provide two separate ball tracks.

Double-Thrusts. These, as shown in Fig. 46E, have two superimposed rows of balls separated by a central grooved race which is rigidly supported in the bearing housing to take the thrust load in both directions. The sizes of these bearings are in accordance with those of the types previously described.

MOUNTING THRUST BEARINGS

The type of thrust bearing fitted to the spindle of, say, a small

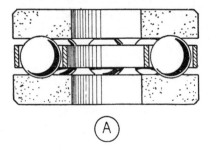

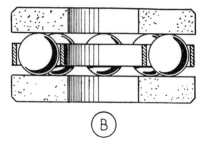

Fig. 46. Types of thrust bearings.

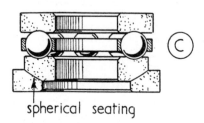

spherical seating

drilling machine is perhaps the form in most general use. This bearing, which is illustrated in Fig. 47, should be installed with the rotating race made a push fit on the machine spindle, and the shoulder against which it abuts must be machined flat and truly at right angles to the long axis of the shaft; in

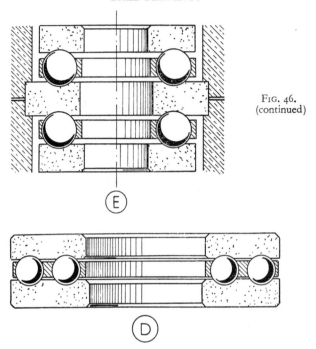

Fig. 46.
(continued)

addition, the shoulder must be of sufficiently large diameter to extend under the ball track and so prevent the race from becoming dished when the thrust load is applied. The bore of the stationary race is, however, made slightly larger so that, when the two ball tracks are located concentrically by the row of balls, this race will stand clear of the rotating shaft. There is no need to fit keys or snugs to prevent the races from rotating, for the friction between the races and their abutments is much greater than that between the balls and their tracks; but if either race tends to rotate or creep, it indicates that the seatings are not truly at right angles to the axis of the rotating shaft.

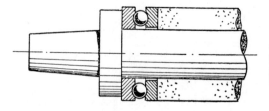

FIG. 47.
Thrust bearings fitted to a machine spindle.

The double thrust bearing, Fig. 46E, is fitted with the large diameter central race securely clamped endwise in a recess, machined in the housing truly at right angles to the shaft axis. The end-play is then taken up by means of a screwed collar fitted to the shaft and securely locked after the adjustment has been made.

Chapter Five

ROLLER BEARINGS

THE roller bearings in common use are of three types: the ordinary journal bearings fitted with one or two rows of parallel, cylindrical rollers: bearings of the well-known Timken pattern having tapered rollers: and a modification of the first type, the so-called needle roller bearings which have long rollers of small diameter.

PARALLEL ROLLER BEARINGS

The line contact between the rollers and their races enables the bearing to carry a heavier load than the corresponding ball bearing, where the balls make point contact only. The rollers fitted to these bearings have a diameter equal to their length and, moreover, their diameter is accurate to within one ten-thousandth and the length to within two ten-thousandths of an inch. As in ball bearings, the rollers are usually mounted in a cage, but for carrying heavy loads the cage may be omitted and the space between the races is then entirely filled with rollers.

The ordinary type of journal bearing shown in Fig. 48A has an inner race formed with two shoulders, but the outer race has a plain, flat roller track. Light, medium and heavy grades of bearings are made, having bores of $\frac{1}{4}$ in. upwards.

For providing end-location in one direction only, the bearing shown in Fig. 48B has a single shoulder formed on the outer race, in addition to the two shoulders on the inner race. Where, however, end-location is required in both directions, the outer race is also furnished with a double shoulder as illustrated in Fig. C.

As in the case of ball bearings, roller bearings are also made of the self-aligning type; here, as depicted in Fig. D, the outer surface of the outer race has a part-spherical form to correspond with the curvature of the special ring-housing in which the bearing is carried. For carrying a heavy load in a restricted space, a double row of rollers may be fitted to the bearing; Fig. E shows the crank pin of a motor-cycle engine equipped with a bearing of this type to accommodate the big end of the connecting rod.

Like their ball bearing counterparts, roller bearings are also designed for carrying purely thrust loads, and, where exceptionally heavy loads of this nature are encountered as in lifting jacks and the steering pivots of heavy vehicles, a thrust bearing may be fitted having a double row of rollers, as illustrated in Fig. F.

MOUNTING ROLLER BEARINGS

The methods of fitting roller bearings are in the main similar to those employed for mounting ball bearings. Care should be taken, however, to ensure that, when the inner and outer races are clamped in place, the rollers lie centrally in the races and are not subjected to end pressure.

The abutment faces should extend as far as the roller tracks, but, when bearings of the light type are fitted, the clamping pressure should not bear on the shoulders of the roller tracks or the free movement of the rollers may be impeded owing to distortion of the race.

In the example illustrated in Fig. 49, two Hoffmann roller bearings are shown fitted to a wheel hub. Both races of the outer bearing have double shoulders so as to provide for end-location of the wheel. The outer races of the two bearings

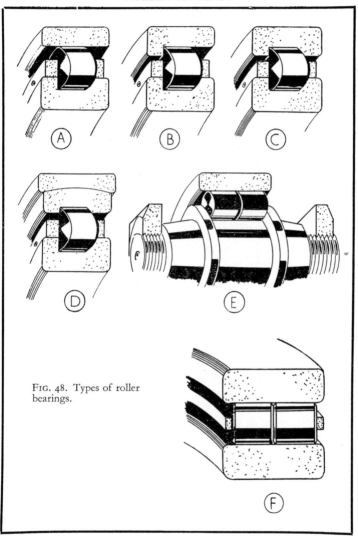

FIG. 48. Types of roller bearings.

are made an interference fit in the hub shell and are clamped endwise. The inner races of both bearings are a push fit on the axle and are secured in place by the large clamp nut.

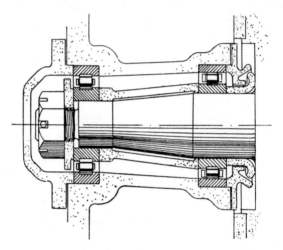

FIG. 49. A wheel hub fitted with roller bearings.

When assembling components on which the roller races have been secured, it is essential to present the parts squarely to one another, otherwise the races may easily be scored.

The inner races of self-aligning roller bearings should be firmly secured to the shaft, but the housings must allow the outer races to slide so that they can take up the correct position in relation to the inner races. Where the end-thrust is inconsiderable, the shaft can be end-located by clamping one outer race in its housing.

The provision of chip shields and devices for retaining the lubricant are just as important for roller bearings as for ball bearings, and these adjuncts can be fitted in a similar manner to that already described.

TAPERED ROLLER BEARINGS

The roller journal bearings so far described are fitted with cylindrical rollers and, as the axis of the rollers is parallel with that of the bearing bore, they have a zero angle of taper and

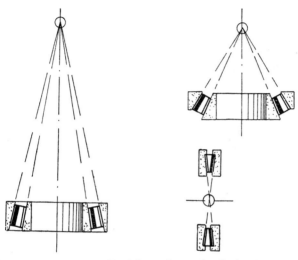

FIG. 50. (*above*). Usual form of tapered roller bearing.
FIG. 51. (*top right*). Bearing specially designed to take thrust.
FIG. 52. (*bottom right*). Section of a tapered roller thrust bearing.

are, therefore, primarily adapted for carrying radial loads. If, however, the rollers and their corresponding roller tracks are tapered, as in the Timken type of bearing, the bearing is then well adapted to carry efficiently both radial and axial loads.

The formation of this taper in the normal type of bearing, suitable for withstanding ordinary journal and thrust loads, is represented in Fig. 50; but should the bearing be required to carry a heavy thrust load in addition to a normal radial load, the angle of taper is then made steeper as shown in Fig. 51.

61

Likewise, when the bearing is designed for thrust loads only, the apex of the angle of taper is made to fall at the central point of the bearing, as represented in Fig. 52.

The ordinary, single-row, tapered roller bearing, shown in Fig. 53, is designed primarily as a journal bearing for use where

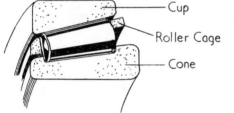

FIG. 53. The components of a tapered roller bearing.

the radial load is in excess of the thrust load, but, as already mentioned, bearings of this pattern are also obtainable with a steep angle of taper suitable for conditions where the thrust is greater than the radial load.

When applied to machine tools, these bearings may conveniently be of the type shown in Fig. 54; here, the outer race

FIG. 54. Bearing with flanged outer race.

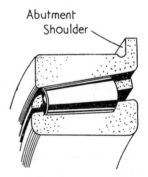

or cup is furnished with a flange for locating the bearing, thus dispensing with the need to machine a shoulder or abutment face within the bearing housing.

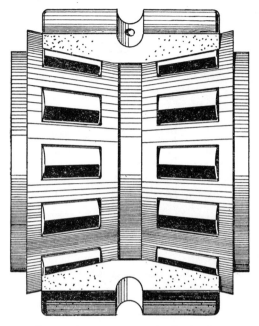

FIG. 55. A two-row tapered roller bearing.

The two-row bearing, illustrated in Fig. 55, is designed to take thrust loads in both directions in addition to the normal journal loading. The cup is formed with a double taper, and two internal races or cones are used; the correct running clearance in the bearing is effected either by fitting a central spacing ring, or an external form of adjustment by means of a threaded collar is provided.

In addition to the types mentioned, there are other varieties of Timken roller bearings suitable for special purposes; for example, a four-row bearing is manufactured to carry extremely heavy loading such as is encountered in the bearings of rolling mills.

lock

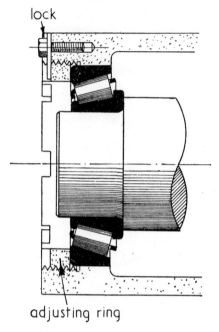

Fig. 56. Method of mounting a Timken bearing.

adjusting ring

MOUNTING TAPERED ROLLER BEARINGS

The British Timken Co. recommend in their instructional handbook that, where the shaft is the rotating member, the inner races or cones should be made a press fit on the shaft and that, as illustrated in Fig. 56, the running clearance in the bearing should be adjusted by means of the stationary cup. On the other hand, where the shaft is stationary, the cones should be a firm sliding fit, and the bearing clearance is set by adjusting the cones themselves; the cups should in this case be made a press fit in their housings.

When fitting either the cup or the cone component of the bearing, it is essential to provide an adequate abutment face or shoulder, and at the same time, as represented in Fig. 57, these shoulders should leave a sufficient amount of the race exposed to enable it when necessary to be withdrawn from its seating.

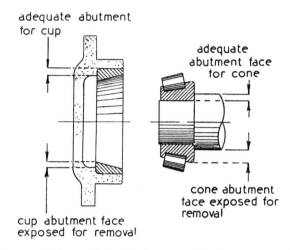

FIG. 57. Fitting a Timken bearing to provide for easy removal.

MACHINE TOOL APPLICATIONS

Where a high degree of accuracy is required when mounting machine tool spindles in Timken bearings, the special precision type of bearings should be fitted.

These are manufactured in two grades: Precision No. 3 and Precision No. 5, and the figures indicate the maximum eccentricity in ten-thousandths of an inch, measured in the unmounted and unloaded bearing while one race revolves and the other remains stationary; but when the bearings are mounted on the machine spindle, the total eccentricity will usually be much reduced.

Needless to say, if bearings of this type are to give satisfactory service they must, in common with other kinds of bearings, be correctly and accurately fitted. As a case in point, when the tapered roller bearings fitted to a lathe mandrel were examined because an irregular finish was given to the work when surfacing, it was found that the bearing seatings on the mandrel had been struck up with a punch to give a semblance of a fit and, in addition, the housings for the cups were roughly machined and the abutment shoulders were inaccurately formed.

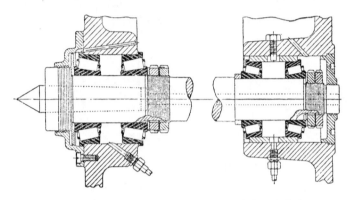

FIG. 58. A rigid form of mounting for a lathe mandrel.

In a lathe mandrel fitted with plain bearings some clearance must be present to allow oil to be retained between the bearing surfaces, but the tapered roller bearing will operate efficiently when this play is eliminated; it is claimed, therefore, that this allows a better finish to be given to the work, particularly where heavy turning operations are undertaken, or high speeds and heavy cuts are employed in conjunction with the use of tungsten carbide tipped tools.

Fig. 58 illustrates the bearing arrangement of a machine spindle designed to afford maximum rigidity. The journal load imposed while the machine is at work is shared between the two bearings at the forward end of the spindle, and, at the same time, the thrust load does not tend to deflect the spindle as it is taken by the front bearing close to the work. The two opposed bearings fitted to the tail end of the mandrel serve still further to enhance the rigidity of the spindle mounting.

It will be seen that the two sets of bearings are adjusted individually by means of collars fitted to the threaded portions

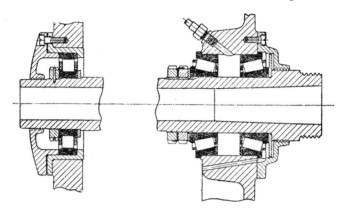

FIG. 59. A lathe mandrel fitted with three bearings.

of the shaft. Furthermore, an adequate lubrication system is provided, and grease seals are fitted at the outer ends of both bearing assemblies.

An alternative design, recommended by Messrs. Timken for the mandrel bearings of a lathe headstock, is illustrated in Fig. 59; the inner bearing cones are a tight fit on the mandrel, and adjustment of the bearing clearance is made in this instance by rotating the threaded collars fitted behind the second bearing of the front pair.

An arrangement suitable for the mandrel of a light lathe is the two-bearing assembly shown in Fig. 60. Here, both radial and thrust loads are taken close to the work, but the bearings are adjusted simultaneously by a common adjusting collar fitted at the rear end of the mandrel. In this design, allowance is made for only small changes in the temperature of the mandrel while in operation, for any marked rise of temperature would cause elongation of the mandrel itself and this, in turn, would give rise to an increase of the bearing clearance.

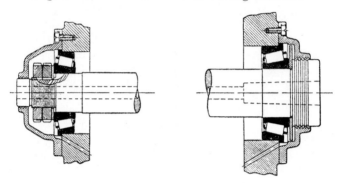

FIG. 60. Bearing arrangement suitable for a light lathe.

NEEDLE ROLLER BEARINGS

These bearings are a modified form of the ordinary parallel roller bearing, but, as the name implies, the rollers are here of greatly increased length in relation to their diameter. In fact, the smallest standard needle rollers manufactured by Messrs. Ransom and Marles have a diameter of only 2.5 mm. and a length of 9.8 mm., whilst in the largest size these dimensions are 4 mm. and 39.8 mm. respectively.

Bearings of this type are particularly suitable for fitting to mechanisms where the motion is ocillatory in character as in valve rockers and universal joint pins, but they also give good service in connection with rotating parts when adopted to save space; for example, when fitted to the big end of a motor-cycle engine connecting rod, they occupy little more room than an ordinary plain bearing bush.

Where the available space is extremely limited, the needle rollers can be fitted directly in contact with the hardened and accurately finished surfaces of the parts themselves, but more often a complete bearing unit is employed consisting of an inner and an outer race between which the rollers are carried. A sectional view of a typical needle roller bearing is given in Fig. 61.

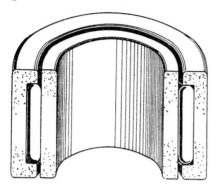

FIG. 61. Sectional drawing of a needle roller bearings.

The rollers themselves have hemispherical ends which run in contact with the shoulders formed in the outer race, but this design allows the rollers to fall out when the inner race is removed during assembly of the bearing in its housing. To overcome this difficulty, Messrs. Ransom and Marles use

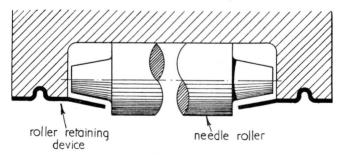

roller retaining device

needle roller

FIG. 62. Method of retaining the needle rollers in position.

trunnion-ended rollers of the form shown in Fig. 62; here, the rollers are retained in the outer race by means of two rings, which ensure that the rollers remain in place during transport and assembly. The bore of the smallest size standard bearing assembly is 12 mm. and the diameter of the outer race 30 mm.

Needle roller bearings have the limitation that they are not designed to withstand thrust, nor should they be used to provide end-location. Apart from this, the methods of mounting and the need for protection are similar to those applied to ordinary ball or roller bearings.

In the example illustrated in Fig. 63, an unhardened steel shaft is carried in a complete bearing unit, and the outer race is retained in its housing by a clamp plate which also serves as a protective cover.

The method of employing a needle roller bearing in a fan mounting is shown in Fig. 64. Here, the fan spindle consists of a hardened and accurately-ground steel shaft on which the rollers run directly. A clamping plate is fitted to retain the outer race in its housing, and the assembly is completed by fitting a cover at the outer end to protect the bearing.

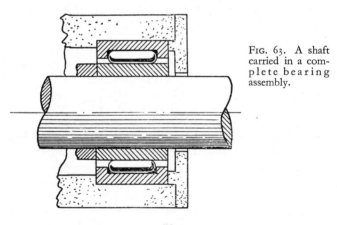

FIG. 63. A shaft carried in a complete bearing assembly.

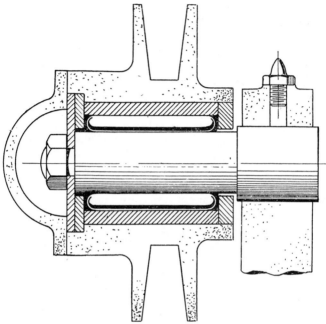

Fig. 64. A pulley mounted on a needle roller bearing.

LUBRICATION OF BALL AND ROLLER BEARINGS

The manufacturers of these bearings point out that the essential functions of a lubricant are: to reduce cage friction; to protect the bearing surfaces from corrosion; and to act as a seal in preventing the ingress of dirt or moisture.

A good quality mineral lubricant is recommended which neither causes gumming nor has any chemical action on the bearings.

Under ordinary running conditions, a thin grease will be found the most suitable type of lubricant, for it provides a ready means of excluding dirt and, when packed into the bearing, it requires renewal only at long intervals. A pressure grease-gun may be used for supplying the lubricant through a suitable form of nipple.

Where the shaft speed is high, say 8,000 r.p.m. and over, a light mineral oil should be employed to lubricate lightly loaded bearings, and a continuous feed of fresh lubricant by means of a wick- or drip-oiler is preferable to using any form of reservoir to maintain the oil supply.

For Timken tapered roller bearings the makers recommend that grease should be used for lubrication where the bearing diameter does not exceed 6 in. nor the revolutions 1,000 per minute, but when high speeds are employed oil should be used instead of grease.

As in the previous instance, the grease can be introduced into the bearing by means of a pressure gun, and, in addition, efficient grease seals should be fitted to ensure that the lubricant is retained and not dissipated by centrifugal force.